# 4th Grade Math
# Volume 3

ISBN: 978-1-939796-84-4

# Table of Contents

# Problem Solving Strategies
## *Inverse Operations*

### *Key Vocabulary*

*inverse operations*

*work backwards*

**Chocolate Smoothie Problem**

Mia made a chocolate smoothie with 12 oz of milk and some chocolate ice cream. She filled four 6-oz glasses and there were 3 oz left over.

How many ounces of chocolate ice cream did she use?

12 + ? = 27

27 − 12 = 15

Mia used 15 oz of ice cream.

**Work backwards using inverse operations to solve problems.**

**Addition and Subtraction are Inverse Operations**

3 + 5 = 8

☐ − ☐ = ☐

**Multiplication and division are inverse operations.**

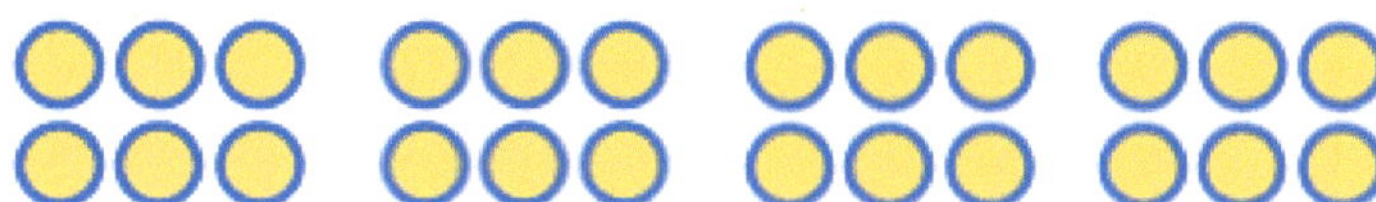

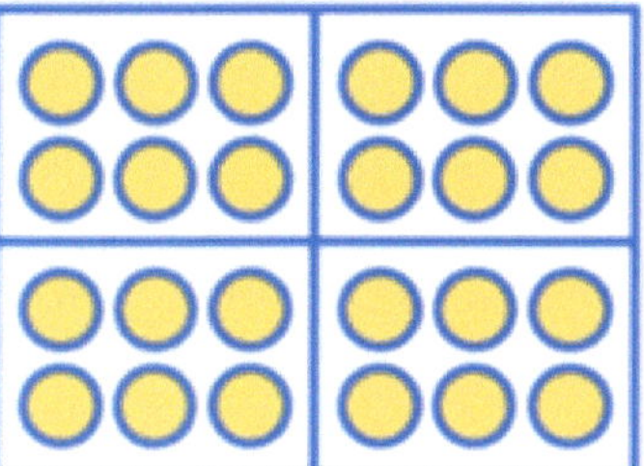

**Use inverse operations to complete these operations.**

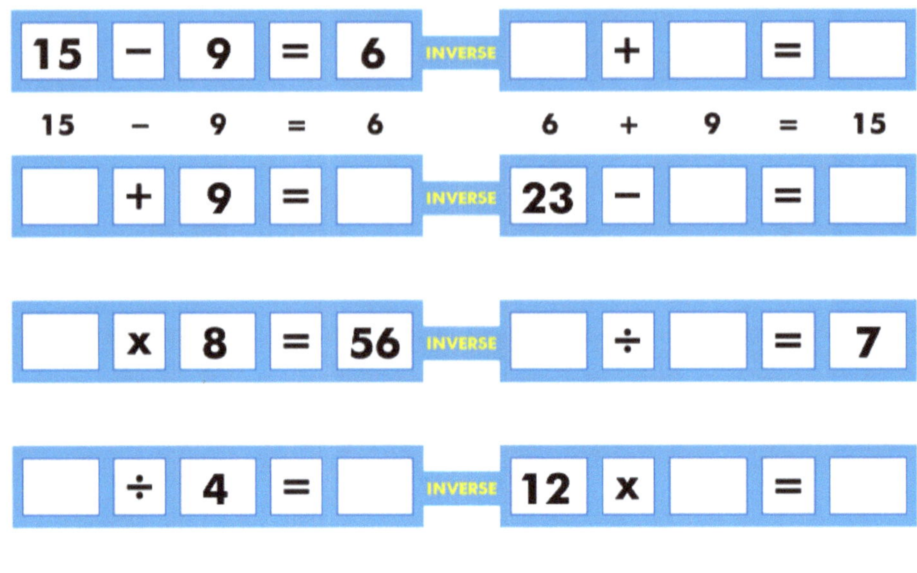


**How old is James?**
Complete the flowchart below by adding x3 and ÷3 to calculating James age.

**James' Dad is 54 years old.  If James starts with his own age, adds 9, then multiplies by 3, he will get to his Dad's age. How old is James? Solve using a flowchart.**

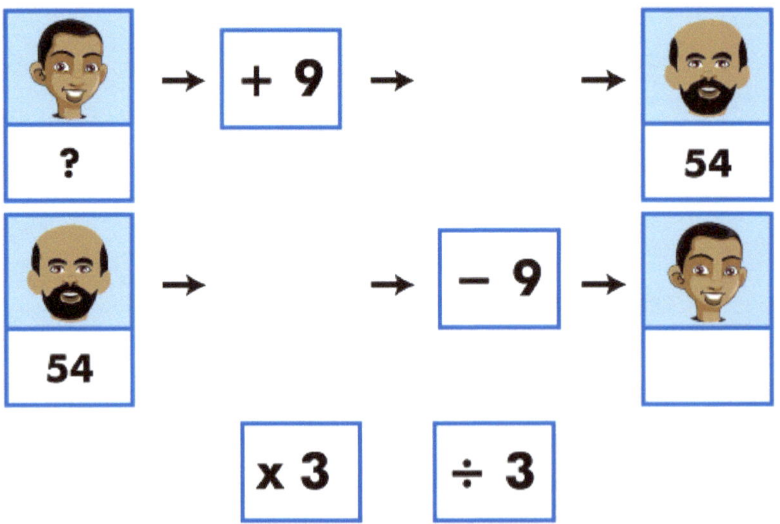

**Work backwards to solve.**

Brian and Jenna are sharing a bowl of pistachio nuts while watching a game of football. In the first half, Brian takes 7 nuts and Jenna takes 3 nuts.  At the start of the second half, Brian takes half of the remaining nuts, while Jenna takes 10 nuts. If there are 15 nuts left for the 4th quarter, how many nuts did they have to begin with?

Name_______________________________________

## Inverse Operations Quiz

**1**   **True or false? Addition and multiplication are inverse operations.**

**2**   **Which equation uses inverse operations to 42 − 11 = 31?**

    **A**   **42 x 11 = 31**

    **B**   **31 + 11 = 42**

    **C**   **42 ÷ 11 = 31**

    **D**   **31 − 11 = 42**

**3**   **Which digit is missing in these inverse equations:**
    **? x 6 = 72      72 ÷ 6 = ?**

**4**   **Ingrid's hamster has pups. A friend takes $\frac{1}{2}$ of the pups, but returns three, leaving Ingrid to care for 15. pups. How many pups were in the litter?**

# Problem Solving Strategies
## *Patterns and Tables*

**Key Vocabulary**

pattern

table

rule

sequence

**Extend a Pattern**
Circle the correct symbol to extend the pattern.

**Find and extend these numerical patterns.**

 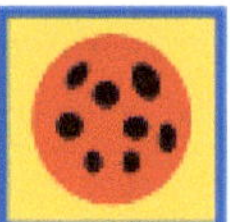 

**1**  60, 80, 100, 120, ______ , ______

**2**  2, 5, 4, 7, 6, ______ , ______

**3**  5, 10, ______ , ______

**Finding and extending a pattern.**
Complete the patterns below.

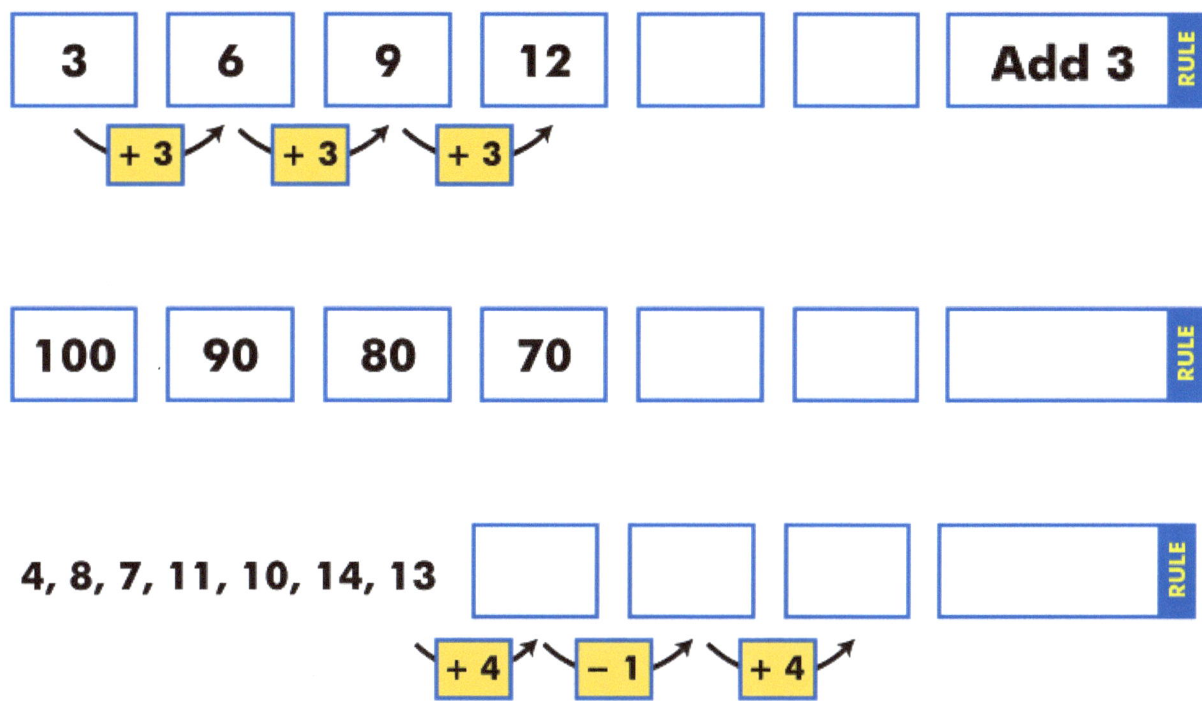

**Organizing a Data Table**
Read the information provided about James reading progress.  Creating a data table is a good strategy to solve this problem.  Discover the data table that has been created and discover how it is used to solve the problem.

**Number of pages read by James each month**

| Month # | 1 | 2 | 3 | 4 | 5 | 6 | 7 | 8 |
|---|---|---|---|---|---|---|---|---|
| Total # Pages | 80 | | | | | | | |

---

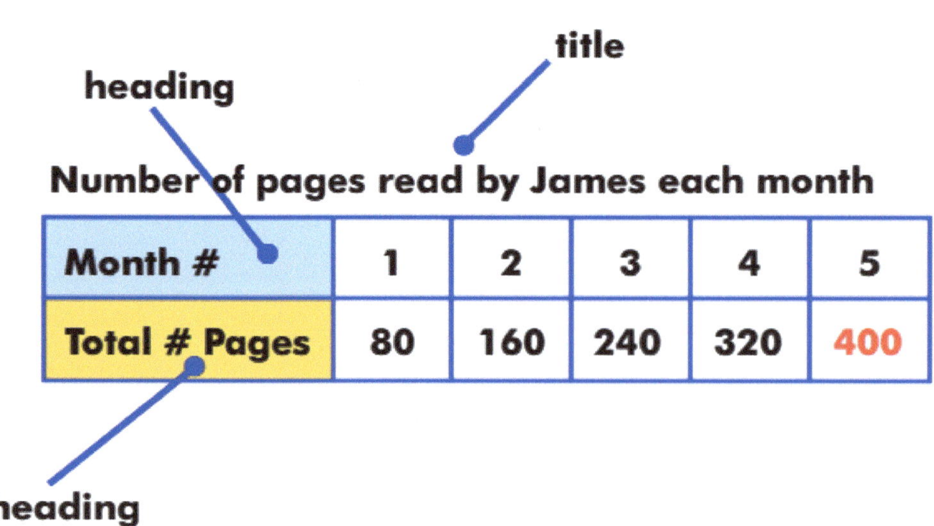


| Month # | 1 | 2 | 3 | 4 | 5 |
|---|---|---|---|---|---|
| Total # Pages | 80 | 160 | 240 | 320 | 400 |

**Tables can help you to organize data**

**Create a table and solve.**
All the elements are provided below.

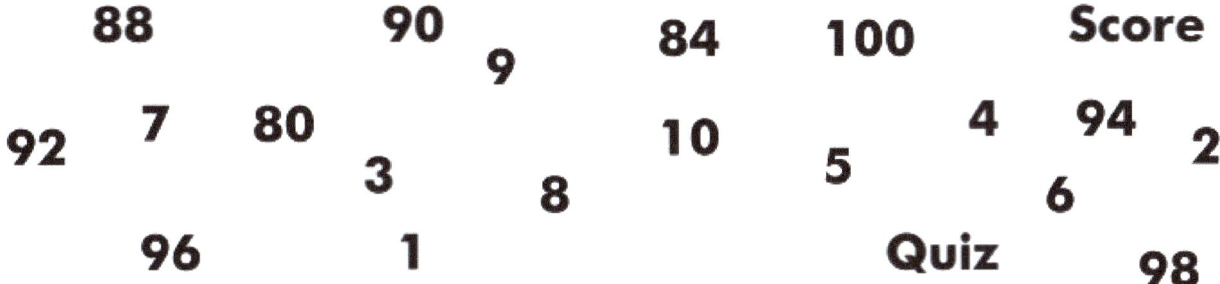

Name_______________________________

## Patterns and Tables Quiz

**1** **True or false, the rule for this pattern is add 4?**
**4, 8, 16, 32...**

**2** **What is the rule for this pattern?  5, 8, 6, 9, 7, 10, 8...**

**A** Add 3

**B** Subtract 2

**C** Add 3, subtract 2

**D** Subtract 3, add 2

| # Boxes | S & H |
|---------|-------|
| 1 | $6 |
| 2 | $11 |
| 3 | $15 |
| 4 | $18 |

Table 1

**3** **What is the shipping and handling (S & H) cost, in dollars, for shipping 5 boxes? (Table 1 above)**

**4** **If Owen earns $18 every week, and saves $8, how many whole weeks will it take him to save enough to buy a $60 digital player?**

# Decimals

**Key Vocabulary**

tenth

hundredth

thousandth

equivalent decimals

How much money  is here?

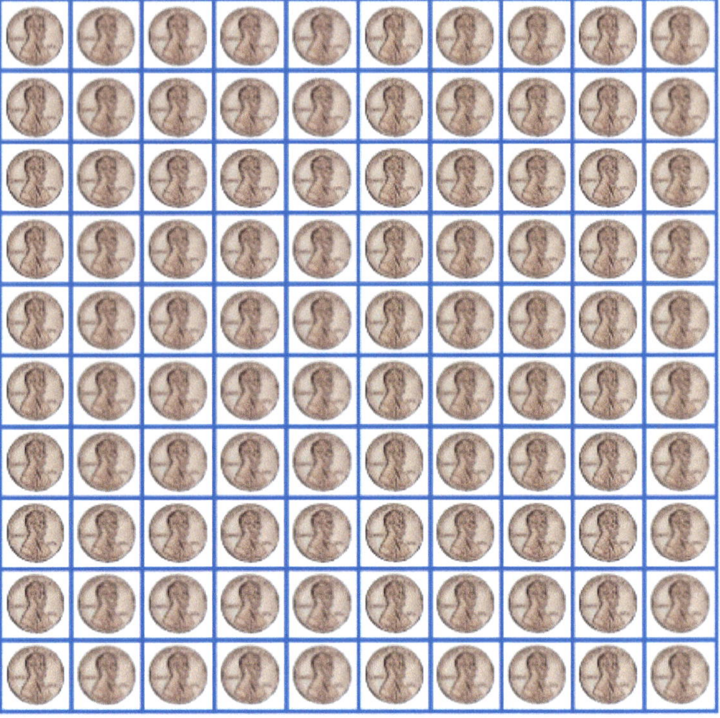

$ ______________


______________¢


How much money is here?

______________¢

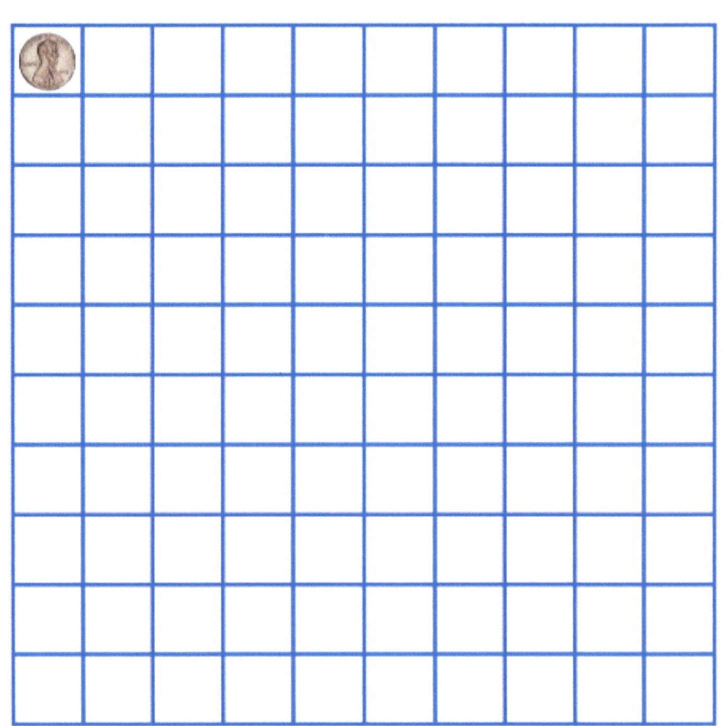

**Write 1¢ as a decimal.**

**Write these coins as decimals.** 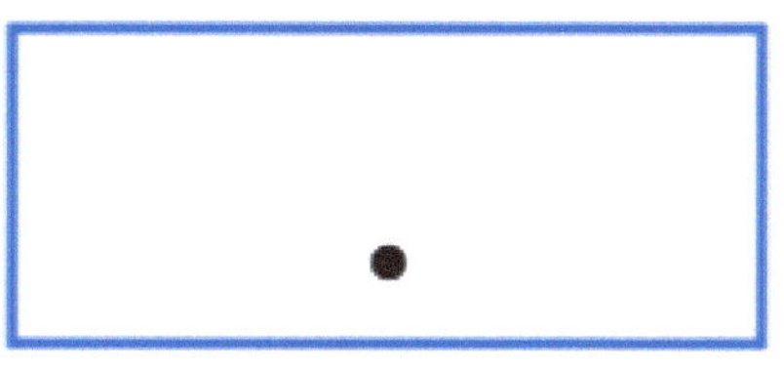

**How do you write this in word form?**     _____ hundredth

_____________

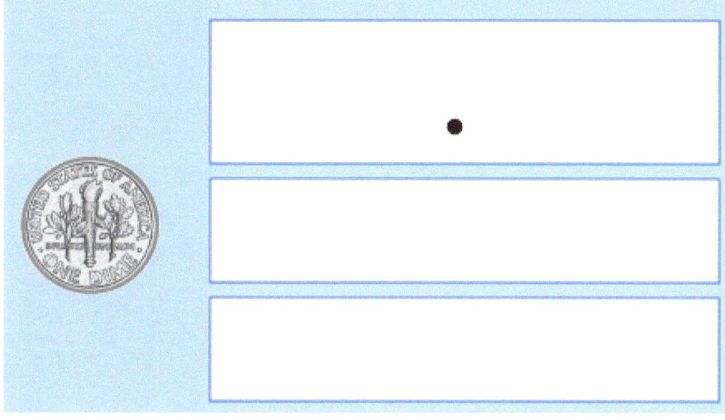

**Write these amounts as decimals.**

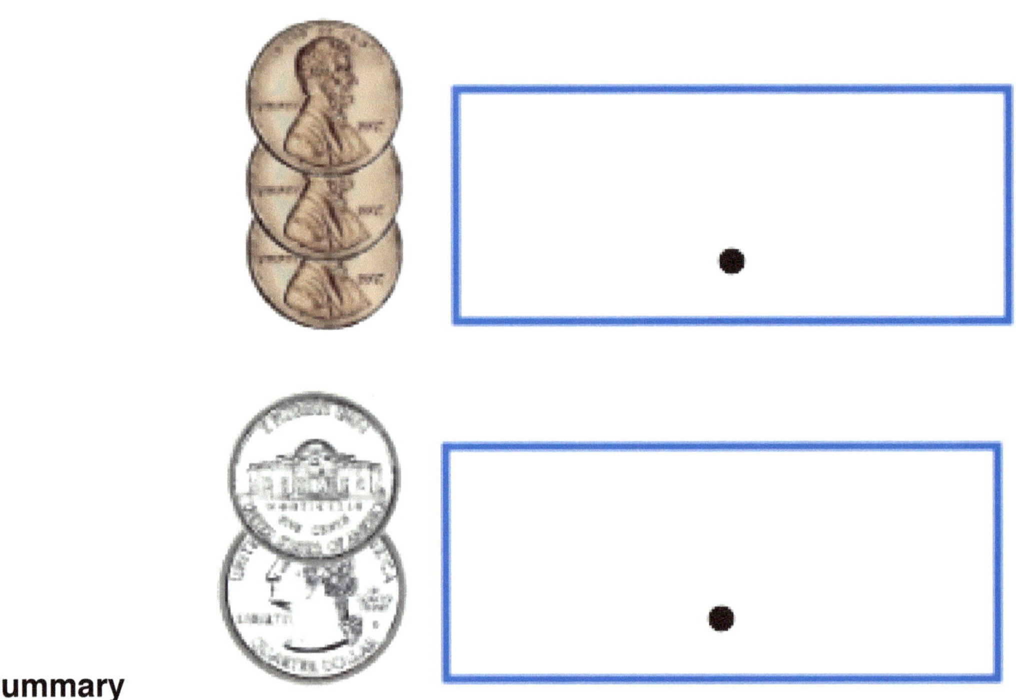

**Summary**

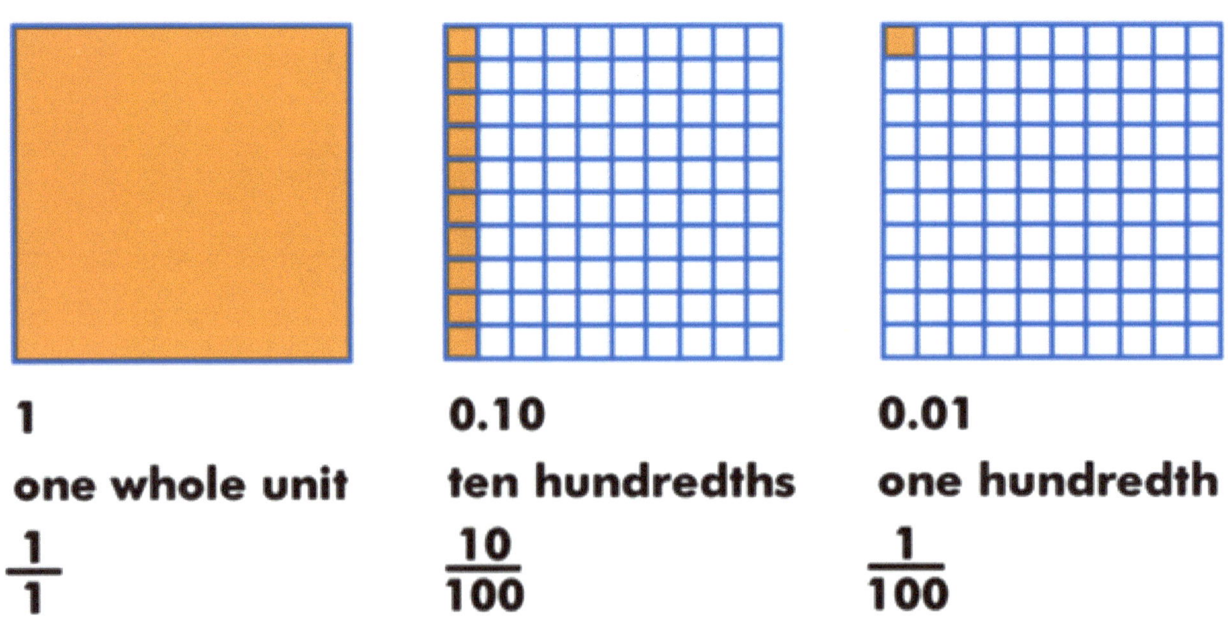

| 1 | 0.10 | 0.01 |
|---|------|------|
| one whole unit | ten hundredths | one hundredth |
| $\dfrac{1}{1}$ | $\dfrac{10}{100}$ | $\dfrac{1}{100}$ |

**Write as a decimal.**

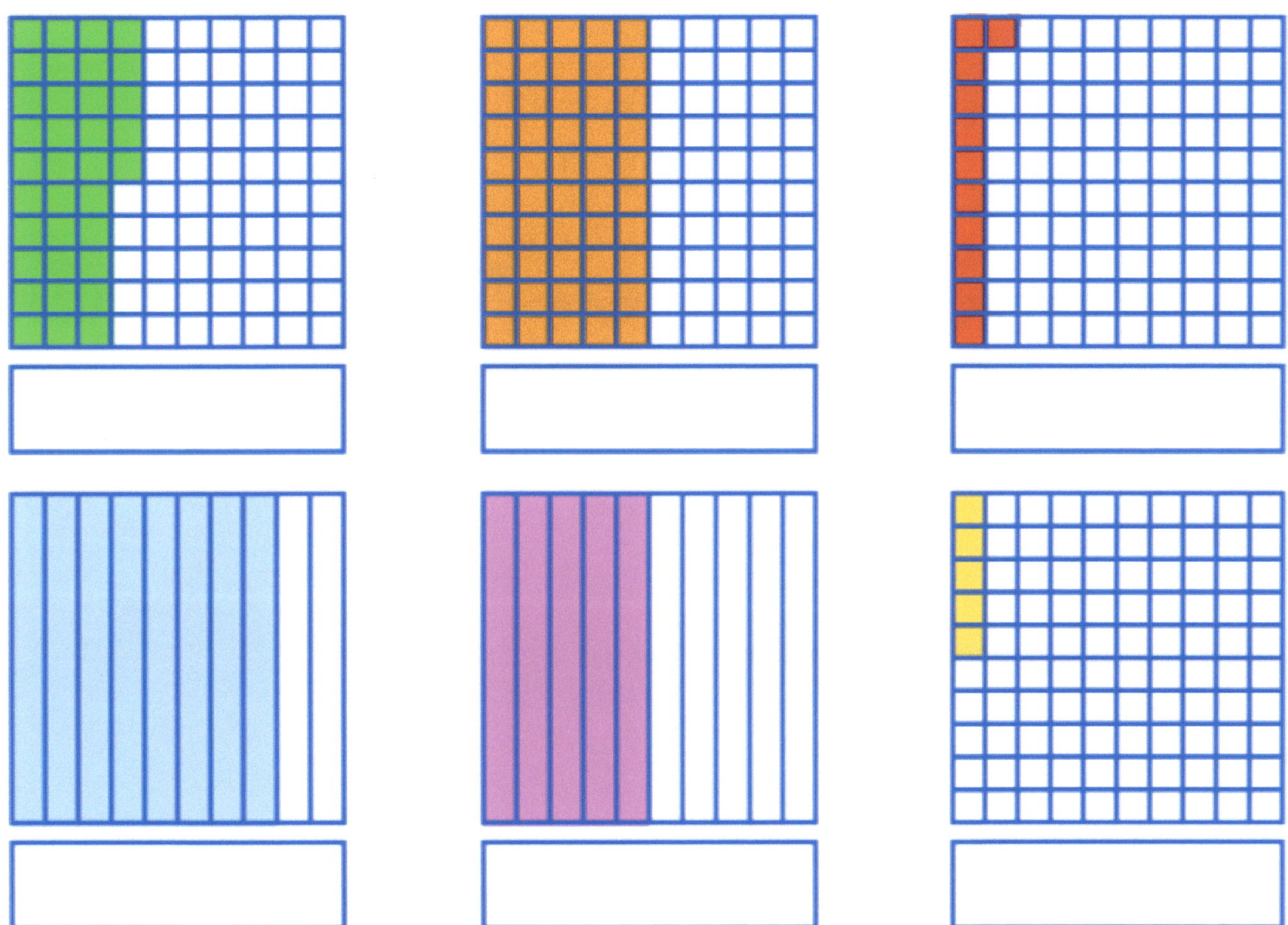

Name___________________________________

## Decimals Quiz

 **1**   **True or false?**  **written as a decimal is 0.01**

**2**   **What is 0.05 in words?**

- **A**   five tenths
- **B**   five hundredths
- **C**   five ones
- **D**   five hundreds

 **3**   **Which is smaller, 0.01 or 0.1?**

**4**   **Which is larger, 0.52 or 0.25?**

# Equivalent Fractions

## Key Vocabulary

numerator

denominator

equivalent fractions

**Who ate the most pizza?**

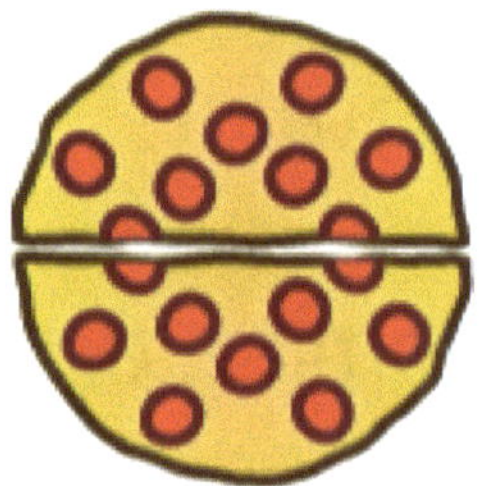



They both ate the same amount. $\frac{2}{4} = \frac{1}{2}$

$\frac{2}{4}$ and $\frac{1}{2}$ are *equivalent fractions*.

# OnBoard Academics Workbook    Grade 4 Mathematics

**Which figures have one half 1/2 shaded?**
Place a check mark int the box if 1/2 shaded and an X if not.
**Circle the equivalent fractions.**

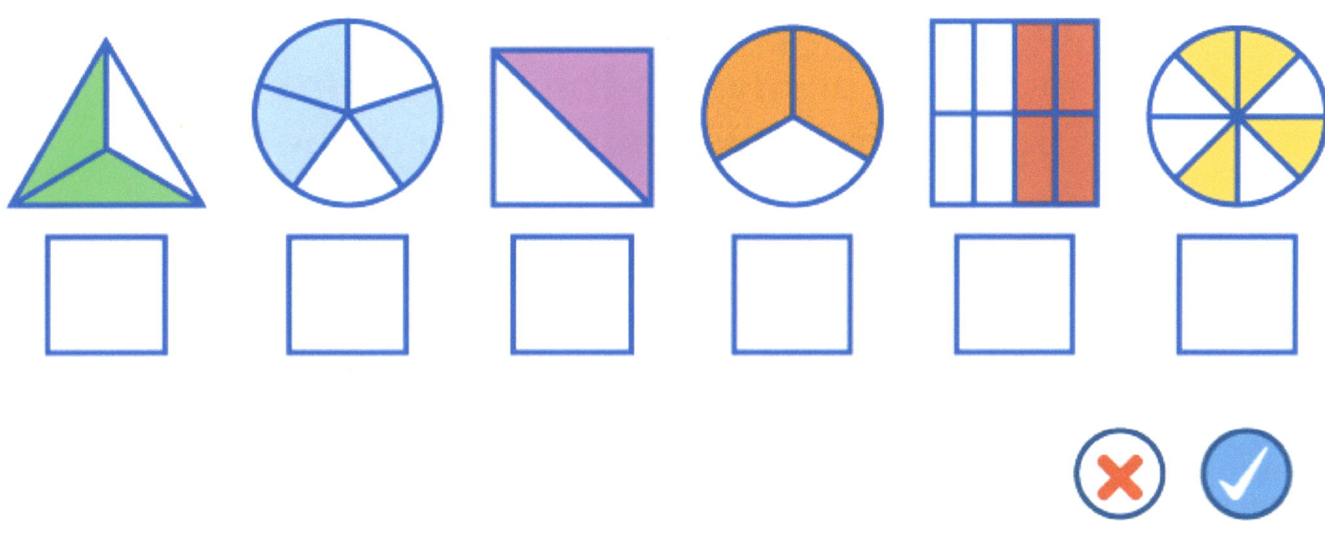

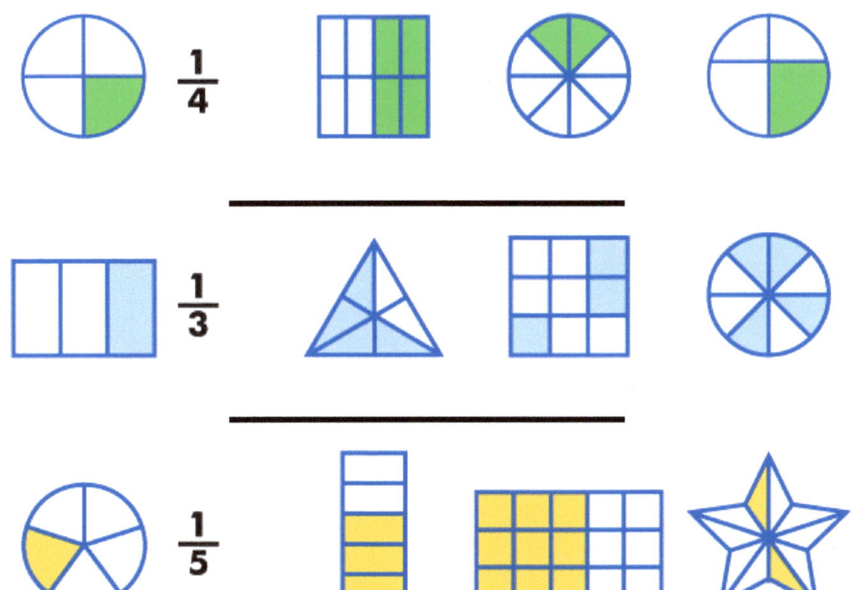

**Finding Equivalent Fractions**
Discover equivalent fractions below.  What is the first fraction multiplied by in example 2 and 3 to make the equivalent fraction?

> **To find an equivalent fraction, multiply (or divide) the numerator and the denominator by the same number.**

x 2

$$\frac{1}{4} = \frac{2}{8}$$

x 2

$$\frac{1}{3} = \frac{3}{9}$$

X ___________

$$\frac{1}{5} = \frac{2}{10}$$

X ___________

**Find Equivalent Fractions**

Fill in the missing boxes using the chart below.

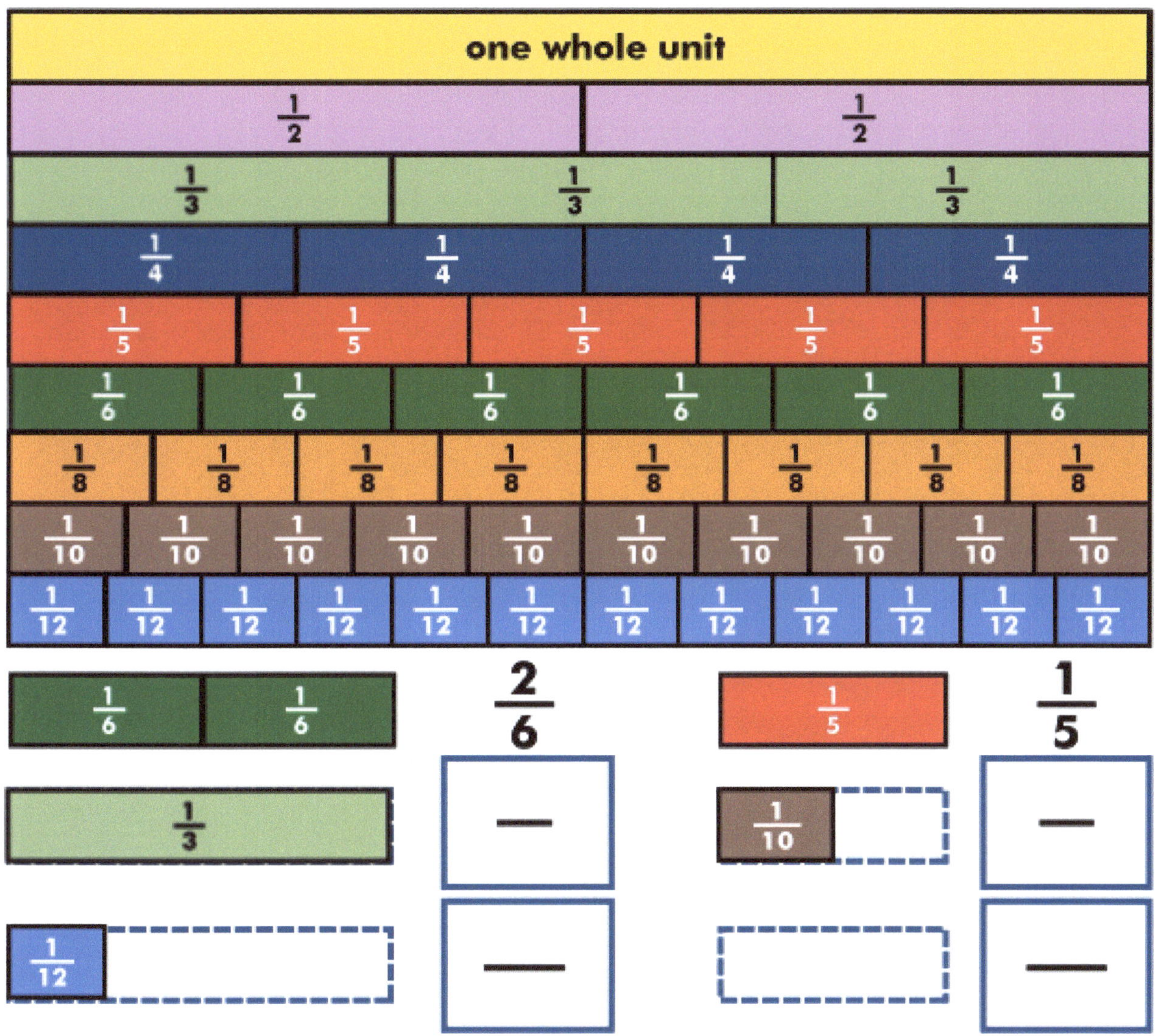

# OnBoard Academics Workbook — Grade 4 Mathematics

**Find Equivalent Fractions**

Fill in the empty boxes using the chart below.

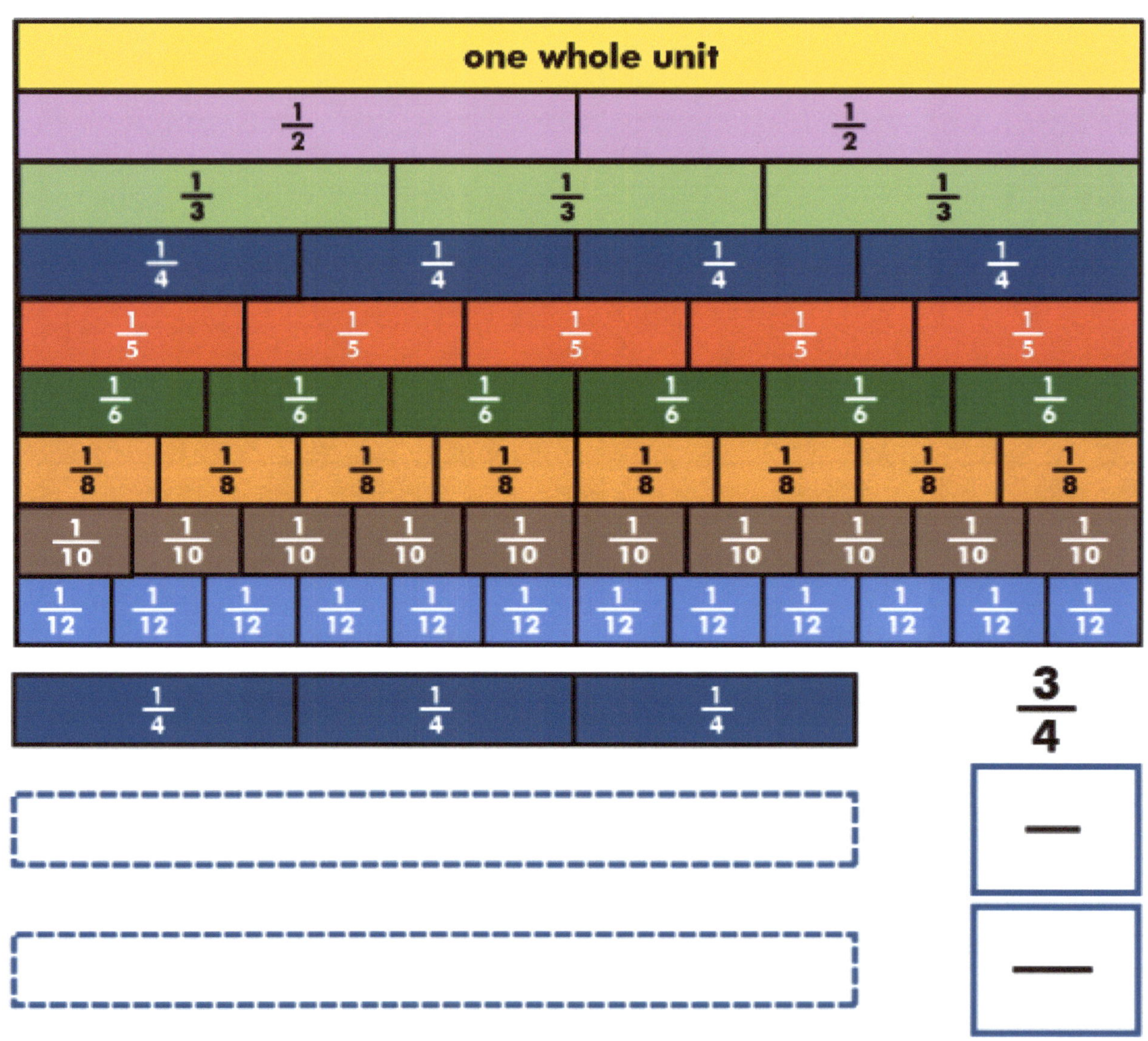

**Find Equivalent Fractions**

Draw the number cards on the number line.  One example has been completed for you.

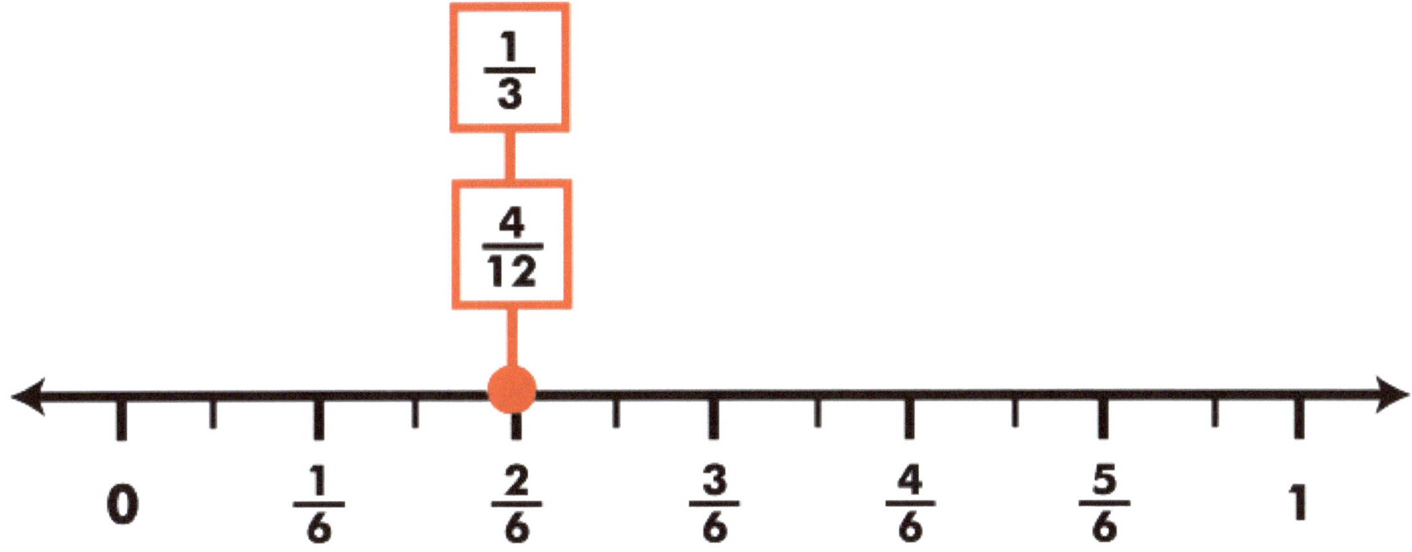

Name______________________________

## Equivalent Fractions Quiz

**1** These two fractions are equivalent, true or false?

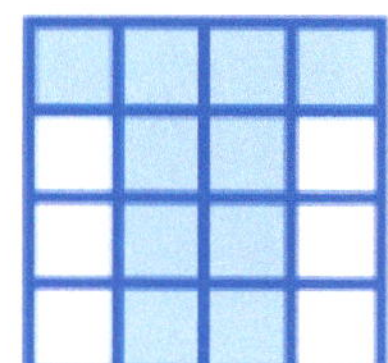 

**2** Which fraction is the odd one out?

  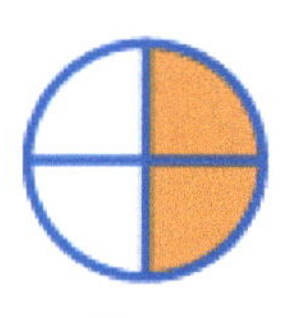 

A          B          C          D

**3** What is the missing numerator?     $\dfrac{5}{10} = \dfrac{?}{40}$

**4** What is the missing denominator?     $\dfrac{3}{8} = \dfrac{9}{?}$

Newburyport, MA 01950

1-800-596-3175

OnBoard Academics employs teachers to make lessons for teachers!  We create and publish a wide range of aligned lessons in math, science and ELA for use on most EdTech devices including whiteboard, tablets, computers and pdfs for printing.

All of our lessons are aligned to the common core, the Next Generation Science Standards and all state standards.

If you like our products please visit our website for information on individual lessons, teachers licenses, building licenses, district licenses and subscriptions.

Thank you for using OnBoard Academic products.

www.ingramcontent.com/pod-product-compliance
Lightning Source LLC
Chambersburg PA
CBHW042136030726
47599CB00002B/491